ESTABLISHMENT AND IMPLEMENTATION OF A TRUSTWORTHINESS PROGRAMME FOR NUCLEAR SECURITY

The following States are Members of the International Atomic Energy Agency:

AFGHANISTAN
ALBANIA
ALGERIA
ANGOLA
ANTIGUA AND BARBUDA
ARGENTINA
ARMENIA
AUSTRALIA
AUSTRIA
AZERBAIJAN
BAHAMAS, THE
BAHRAIN
BANGLADESH
BARBADOS
BELARUS
BELGIUM
BELIZE
BENIN
BOLIVIA, PLURINATIONAL
  STATE OF
BOSNIA AND HERZEGOVINA
BOTSWANA
BRAZIL
BRUNEI DARUSSALAM
BULGARIA
BURKINA FASO
BURUNDI
CABO VERDE
CAMBODIA
CAMEROON
CANADA
CENTRAL AFRICAN
  REPUBLIC
CHAD
CHILE
CHINA
COLOMBIA
COMOROS
CONGO
COOK ISLANDS
COSTA RICA
CÔTE D'IVOIRE
CROATIA
CUBA
CYPRUS
CZECH REPUBLIC
DEMOCRATIC REPUBLIC
  OF THE CONGO
DENMARK
DJIBOUTI
DOMINICA
DOMINICAN REPUBLIC
ECUADOR
EGYPT
EL SALVADOR
ERITREA
ESTONIA
ESWATINI
ETHIOPIA
FIJI
FINLAND
FRANCE
GABON
GAMBIA, THE

GEORGIA
GERMANY
GHANA
GREECE
GRENADA
GUATEMALA
GUINEA
GUYANA
HAITI
HOLY SEE
HONDURAS
HUNGARY
ICELAND
INDIA
INDONESIA
IRAN, ISLAMIC REPUBLIC OF
IRAQ
IRELAND
ISRAEL
ITALY
JAMAICA
JAPAN
JORDAN
KAZAKHSTAN
KENYA
KOREA, REPUBLIC OF
KUWAIT
KYRGYZSTAN
LAO PEOPLE'S DEMOCRATIC
  REPUBLIC
LATVIA
LEBANON
LESOTHO
LIBERIA
LIBYA
LIECHTENSTEIN
LITHUANIA
LUXEMBOURG
MADAGASCAR
MALAWI
MALAYSIA
MALI
MALTA
MARSHALL ISLANDS
MAURITANIA
MAURITIUS
MEXICO
MONACO
MONGOLIA
MONTENEGRO
MOROCCO
MOZAMBIQUE
MYANMAR
NAMIBIA
NEPAL
NETHERLANDS,
  KINGDOM OF THE
NEW ZEALAND
NICARAGUA
NIGER
NIGERIA
NORTH MACEDONIA
NORWAY
OMAN

PAKISTAN
PALAU
PANAMA
PAPUA NEW GUINEA
PARAGUAY
PERU
PHILIPPINES
POLAND
PORTUGAL
QATAR
REPUBLIC OF MOLDOVA
ROMANIA
RUSSIAN FEDERATION
RWANDA
SAINT KITTS AND NEVIS
SAINT LUCIA
SAINT VINCENT AND
  THE GRENADINES
SAMOA
SAN MARINO
SAUDI ARABIA
SENEGAL
SERBIA
SEYCHELLES
SIERRA LEONE
SINGAPORE
SLOVAKIA
SLOVENIA
SOMALIA
SOUTH AFRICA
SPAIN
SRI LANKA
SUDAN
SWEDEN
SWITZERLAND
SYRIAN ARAB REPUBLIC
TAJIKISTAN
THAILAND
TOGO
TONGA
TRINIDAD AND TOBAGO
TUNISIA
TÜRKİYE
TURKMENISTAN
UGANDA
UKRAINE
UNITED ARAB EMIRATES
UNITED KINGDOM OF
  GREAT BRITAIN AND
  NORTHERN IRELAND
UNITED REPUBLIC OF TANZANIA
UNITED STATES OF AMERICA
URUGUAY
UZBEKISTAN
VANUATU
VENEZUELA, BOLIVARIAN
  REPUBLIC OF
VIET NAM
YEMEN
ZAMBIA
ZIMBABWE

The Agency's Statute was approved on 23 October 1956 by the Conference on the Statute of the IAEA held at United Nations Headquarters, New York; it entered into force on 29 July 1957. The Headquarters of the Agency are situated in Vienna. Its principal objective is "to accelerate and enlarge the contribution of atomic energy to peace, health and prosperity throughout the world".

ESTABLISHMENT AND IMPLEMENTATION OF A TRUSTWORTHINESS PROGRAMME FOR NUCLEAR SECURITY

TECHNICAL GUIDANCE

INTERNATIONAL ATOMIC ENERGY AGENCY
VIENNA, 2026

IAEA Library Cataloguing in Publication Data

Names: International Atomic Energy Agency.
Title: Establishment and implementation of a trustworthiness programme for nuclear security / International Atomic Energy Agency.
Description: Vienna : International Atomic Energy Agency, 2026. | Series: IAEA nuclear security series, ISSN 1816–9317 ; no. 51-T | Includes bibliographical references.
Identifiers: IAEAL 25-01784 | ISBN 978-92-0-118925-7 (paperback : alk. paper) | ISBN 978-92-0-118825-0 (pdf) | ISBN 978-92-0-118725-3 (epub)
Subjects: Nuclear facilities — Security measures. | Nuclear facilities — Safety regulations. | Nuclear industry — Materials management. | Nuclear nonproliferation.
Classification: UDC 341.67 | STI/PUB/2126

FOREWORD

by Rafael Mariano Grossi
Director General

The IAEA Nuclear Security Series provides international consensus guidance on all aspects of nuclear security to support States as they work to fulfil their responsibility for nuclear security. The IAEA establishes and maintains this guidance as part of its central role in providing nuclear security related international support and coordination.

The IAEA Nuclear Security Series was launched in 2006 and is continuously updated by the IAEA in cooperation with experts from Member States. As Director General, I am committed to ensuring that the IAEA maintains and improves upon this integrated, comprehensive and consistent set of up to date, user friendly and fit for purpose security guidance publications of high quality. The proper application of this guidance in the use of nuclear science and technology should offer a high level of nuclear security and provide the confidence necessary to allow for the ongoing use of nuclear technology for the benefit of all.

Nuclear security is a national responsibility. The IAEA Nuclear Security Series complements international legal instruments on nuclear security and serves as a global reference to help parties meet their obligations. While the security guidance is not legally binding on Member States, it is widely applied. It has become an indispensable reference point and a common denominator for the vast majority of Member States that have adopted this guidance for use in national regulations to enhance nuclear security in nuclear power generation, research reactors and fuel cycle facilities as well as in nuclear applications in medicine, industry, agriculture and research.

The guidance provided in the IAEA Nuclear Security Series is based on the practical experience of its Member States and produced through international consensus. The involvement of the members of the Nuclear Security Guidance Committee and others is particularly important, and I am grateful to all those who contribute their knowledge and expertise to this endeavour.

The IAEA also uses the guidance in the IAEA Nuclear Security Series when it assists Member States through its review missions and advisory services. This helps Member States in the application of this guidance and enables valuable experience and insight to be shared. Feedback from these missions and services, and lessons identified from events and experience in the use and application of security guidance, are taken into account during their periodic revision.

I believe the guidance provided in the IAEA Nuclear Security Series and its application make an invaluable contribution to ensuring a high level of nuclear security in the use of nuclear technology. I encourage all Member States to promote and apply this guidance, and to work with the IAEA to uphold its quality now and in the future.

CONTENTS

1. INTRODUCTION

BACKGROUND

1.1. There is a need for States to establish and maintain a trustworthiness programme as a part of their nuclear security regime. Such a programme is intended to determine the trustworthiness of persons with authorized access to nuclear material and nuclear facilities; other radioactive material and associated facilities and activities; sensitive information and sensitive information assets, including those that may be accessed remotely; the transport of nuclear and other radioactive material; and nuclear security activities in the areas of detection and response.

1.2. Recommendations on the establishment of measures to ensure the trustworthiness of personnel in order to protect against insider adversaries are provided in the following IAEA Nuclear Security Series publications: No. 13, Nuclear Security Recommendations on Physical Protection of Nuclear Material and Nuclear Facilities (INFCIRC/225/Rev. 5) [1]; No. 14, Nuclear Security Recommendations on Radioactive Material and Associated Facilities [2]; and No. 15, Nuclear Security Recommendations on Nuclear and Other Radioactive Material out of Regulatory Control [3].

1.3. This publication provides technical guidance on how to effectively design and implement trustworthiness programmes for nuclear security.

1.4. The following publications also address the trustworthiness of personnel as applied to nuclear security:

— The Amendment to the Convention on the Physical Protection of Nuclear Material [4];
— IAEA Nuclear Security Series No. 7, Nuclear Security Culture [5];
— IAEA Nuclear Security Series No. 8-G (Rev. 1), Preventive and Protective Measures Against Insider Threats [6];
— IAEA Nuclear Security Series No. 21, Nuclear Security Systems and Measures for the Detection of Nuclear and Other Radioactive Material out of Regulatory Control [7];
— IAEA Nuclear Security Series No. 23-G (Rev. 1), Security of Sensitive Information for Nuclear Security [8];
— IAEA Nuclear Security Series No. 25-G, Use of Nuclear Material Accounting and Control for Nuclear Security Purposes at Facilities [9];

— IAEA Nuclear Security Series No. 26-G, Security of Nuclear Material in Transport [10];
— IAEA Nuclear Security Series No. 27-G, Physical Protection of Nuclear Material and Nuclear Facilities (Implementation of INFCIRC/225/Revision 5) [11];
— IAEA Nuclear Security Series No. 36-G, Preventive Measures for Nuclear and Other Radioactive Material out of Regulatory Control [12];
— IAEA Nuclear Security Series No. 43-T, Security Management of Radioactive Material in Use and Storage and of Associated Facilities [13].

OBJECTIVE

1.5. The objective of this publication is to provide guidance on establishing and implementing a trustworthiness programme for nuclear security based on preventive and protective measures against the threats posed by insiders.

1.6. This publication is intended to be used by competent authorities, licensees (operators) and other organizations with responsibility for the security of nuclear material and nuclear facilities or for other radioactive material and associated facilities and activities. It aims to support the security of such material in use, storage or transport and the security of sensitive information and sensitive information assets, including information that can be accessed remotely. This publication could also be useful for stakeholders (e.g. law enforcement, customs officials, border security) involved in the detection of and response to nuclear and other radioactive material out of regulatory control or other nuclear security events.

SCOPE

1.7. This publication describes methods that can be used to perform assessments and trustworthiness determinations for individuals and for monitoring their ongoing reliability and integrity. This publication provides information on granting access authorization to nuclear material and nuclear facilities, other radioactive material and associated facilities and activities, and sensitive information and sensitive information assets, including information that can be accessed remotely. It also demonstrates how a trustworthiness programme is intended to be a part of defence in depth for nuclear security.

1.8. This publication does not describe in detail how a State should implement trustworthiness measures or gather information to support trustworthiness assessments, since the implementation of these measures depends on national legislative and regulatory requirements. This publication does not prescribe a single approach to implementing a trustworthiness programme, but presents guidance based on examples of good practice. Furthermore, trustworthiness programmes may vary from one State to another. This publication assumes that the competent authority is responsible for both the trustworthiness assessment and the trustworthiness determination; however, in some States the licensee might be responsible for the trustworthiness assessment, the trustworthiness determination, or both. The technical guidance provided in this publication should be understood as applying to assessments and determinations conducted by either competent authorities or licensees.

STRUCTURE

1.9. Section 2 provides an overview of a trustworthiness programme, outlining the benefits, the methodologies for determining trustworthiness and the interface with nuclear security culture programmes. Section 3 examines the roles of the State, designated competent authorities for trustworthiness assessments and determinations, licensees, individuals and other stakeholders. Section 4 describes the implementation of a trustworthiness programme using a graded approach. Section 5 addresses the establishment of criteria for conducting trustworthiness assessments and making trustworthiness determinations. Section 6 describes processes for the implementation of a trustworthiness programme. Section 7 describes the maintenance of records for a trustworthiness programme. Section 8 describes the process for granting, suspending, revoking and reinstating access authorizations. Section 9 provides information on the protection of individuals subject to a trustworthiness programme. Section 10 describes trustworthiness programme reporting criteria. Section 11 describes audits and inspections of a trustworthiness programme.

2. OVERVIEW OF A TRUSTWORTHINESS PROGRAMME FOR NUCLEAR SECURITY

2.1. A trustworthiness programme should be mandated in a State's legislative and regulatory framework, as stated in para. 3.14 of Ref. [1]:

"Taking into consideration State laws, regulations, or policies regarding personal privacy and job requirements, the State should determine the trustworthiness policy intended to identify the circumstances in which a trustworthiness determination is required and how it is made, using a *graded approach*. In implementing this policy, the State should ensure that processes are in place to determine the trustworthiness of persons with authorized access to sensitive information or, as applicable, to *nuclear material or nuclear facilities*."

2.2. Trustworthiness is the characteristic of an individual who behaves consistently — in accordance with cultural, ethical and legal standards — with reliability and integrity, particularly in situations where the individual is not being observed (or is not aware of being observed).

2.3. A trustworthiness assessment is the process of gathering appropriate information (see para. 4.5) on the basis of the level of access an individual needs to nuclear material and nuclear facilities, other radioactive material and associated facilities and activities, or sensitive information and sensitive information assets. The trustworthiness assessment results in either a positive or negative trustworthiness determination of an individual's behavioural indicators of reliability and integrity. A trustworthiness determination is based on information known at a particular point in time and is supported by ongoing assessments (e.g. through self-reporting or due to changes in circumstances), oversight by managers and security departments, a behaviour observation programme, a fitness for duty programme and a nuclear security culture that encourages appropriate behaviour and the reporting of concerns (e.g. zero tolerance of harassment and bullying by peers or managers, reporting hotlines).

2.4. A trustworthiness assessment is a preventive measure intended to reduce the insider threat. Trustworthiness assessments to identify undesirable characteristics can be difficult to undertake because the intent and motivation (e.g. money, ideology, coercion, ego, revenge) of an insider adversary are not directly observable. Although the behavioural patterns of insider adversaries can vary significantly, aberrant behaviour can be easily observed. The observation of such behaviour is an important element of effective trustworthiness assessments, providing evidence either of a potential concern or, through further investigation, of contextually appropriate behaviour.

2.5. A trustworthiness programme consists of a set of processes for collecting information on individuals, assessing this information against defined criteria and determining whether trustworthiness has been established for these individuals. Trustworthiness programmes include initial assessments, periodic assessment, routine reviews, ongoing monitoring, special reviews, employee reporting and cooperation between departments or organizations (e.g. security departments, human resources departments, medical departments, supply chain organizations). Making a trustworthiness determination before granting an individual access to secure locations, sensitive information or sensitive information assets may reduce the potential risk from insider adversaries. However, this should supplement other control measures to increase effectiveness.

2.6. The continuous implementation of a trustworthiness programme for nuclear security can serve as a deterrent to the attempt of criminal or intentional unauthorized acts by insider adversaries, since individuals are mindful of being observed and modify their behaviour to meet the expectations of the licensee. For this reason, if individuals demonstrate aberrant behaviour after receiving a positive trustworthiness determination, it may provide a possible indication of their intention to undertake a criminal or intentional unauthorized act (see para. 5.5). Once such behaviour is reported, it is good practice to act in accordance with predefined criteria, such as revoking or suspending an individual's access until an inquiry can be undertaken to determine the next steps, as appropriate (e.g. return to work, rehabilitation, termination of employment).

2.7. While individuals tend to be consistent in their behaviour over time, behavioural patterns can change when an individual experiences significant or traumatic incidents. For this reason, a graded approach, established on the basis of risk assessments and available resources, should be used to determine the frequency and rigour of the scheduled trustworthiness assessments. Effective threat mitigation should include additional trustworthiness assessments on an 'as needed' basis, for example after a significant or traumatic incident or observation of aberrant behaviour. Continued trustworthiness assessments during employment could identify individuals whose behaviour and characteristics negatively change over time.

2.8. Upon termination of employment, an individual's access to the premises and assets of the licensee should be immediately discontinued, including access to sensitive information and sensitive information assets. Termination procedures should include using a non-disclosure agreement to protect sensitive information and changing encryption keys, passwords and access codes [6]. It is good practice for the licensee or competent authority to recover and account for all official property (e.g. passes, including those for vehicles; uniforms; branded workwear).

2.9. A trustworthiness programme is intended to help identify precursory insider adversary behavioural patterns. An effective trustworthiness programme also focuses on identifying and addressing cultural or organizational factors that can reinforce such behaviour.

METHODOLOGIES FOR DETERMINING TRUSTWORTHINESS

2.10. The methodology used to determine the trustworthiness of individuals varies depending on State laws, regulations or policies concerning personal privacy, and the job position. However, the methods used by licensees or competent authorities for developing a trustworthiness programme for nuclear security generally include the following elements:

(a) Establishing a policy for trustworthiness assessments using a graded approach that is commensurate with the access, authority and knowledge that the individual needs in relation to the facility and its assets. Access to sensitive information at the facility or while working remotely should also be considered.
(b) Developing objective criteria to form the basis for trustworthiness determinations.
(c) Identifying a range of aberrant behaviour that may demonstrate an increased risk of an insider becoming an insider adversary.
(d) Planning trustworthiness assessments for individuals both before and during employment, which include the following activities:
　(i) Gathering information to determine whether an individual has exhibited any aberrant behaviour.
　(ii) Validating the information gathered.
　(iii) Assessing the information against established aberrant behaviour criteria, making a trustworthiness determination for the individual through the use of a decision making framework that allows previously collected information (e.g. criminal records, previous negative trustworthiness determinations) to be considered for its current relevance. For example, behaviour could be assessed on the

basis of how long ago it arose, the seriousness of the issue or the age of the individual at the time at which it occurred.

(iv) Communicating and documenting an individual's trustworthiness determination.

(v) Including a process by which an individual who has received a negative trustworthiness determination can appeal against the result, if permitted by State laws.

INTERFACES OF TRUSTWORTHINESS PROGRAMMES WITH NUCLEAR SECURITY CULTURE PROGRAMMES

2.11. Reference [5] defines the concepts, characteristics and indicators of a nuclear security culture while also providing a model of an effective nuclear security culture and describing the roles and responsibilities of institutions and individuals. IAEA Nuclear Security Series No. 28-T, Self-assessment of Nuclear Security Culture in Facilities and Activities [14], provides guidance on a comprehensive methodology for evaluating nuclear security culture in practice, and IAEA Nuclear Security Series No. 38-T, Enhancing Nuclear Security Culture in Organizations Associated with Nuclear and Other Radioactive Material [15], provides practical guidance on how to implement a systematic approach to enhancing nuclear security culture.

2.12. A robust nuclear security culture is an important element to counter both insider and external threats. The overall objective of nuclear security culture is to establish an organizational culture in which individuals willingly view nuclear security as their personal responsibility. Reference [5] defines 30 observable characteristics that can be used as indicators of an organization's nuclear security culture. Leadership behaviour that can foster effective nuclear security includes management oversight, well developed management systems that prioritize security, and an established behaviour observation programme to assess the trustworthiness of individuals on a continual basis. Examples of positive personnel behaviour that can contribute to effective nuclear security include professional conduct and adherence to procedures, performing tasks as assigned, correct handling of sensitive information and maintaining an awareness of potential safety and security concerns.

2.13. The effectiveness of a trustworthiness programme for nuclear security is enhanced not only by initial and ongoing trustworthiness assessments of individuals but also by the establishment of a robust nuclear security culture.

2.14. A strong nuclear security culture strengthens an individual's trustworthiness through a system in which peers and management respond appropriately to aberrant behaviour and reinforce behaviour that is consistent with nuclear security. Licensees should have employment policies that promote and reward positive traits and desired behaviour, such as respect for equality and professional conduct, while discouraging and penalizing inappropriate behaviour (e.g. harassment, bullying) in an effort to avoid individuals feeling marginalized, demotivated, disgruntled and/or disaffected. Assistance programmes and reporting hotlines can help individuals to raise concerns, seek help and improve the nuclear security culture through direct or anonymous reporting.

INTERFACES WITH SAFETY PROGRAMMES

2.15. Human performance and human reliability are of vital importance in operating facilities and conducting activities in a safe, secure and effective manner. Safety programmes and human reliability programmes include measures to monitor human performance and reliability and can be used in the context of a trustworthiness programme. For example, requirements for medical and psychological fitness for duty (e.g. testing for drug or alcohol abuse) of personnel are established in IAEA Safety Standards Series No. SSR-2/2 (Rev. 1), Safety of Nuclear Power Plants: Commissioning and Operation [16], and recommendations on establishing and implementing policies to address these requirements are provided in IAEA Safety Standards Series Nos SSG-75, Recruitment, Qualification and Training of Personnel for Nuclear Power Plants [17], and SSG-84, The Operating Organization and the Recruitment, Training and Qualification of Personnel for Research Reactors [18]. Reference [19] describes the assessment of competencies.

3. ROLES AND RESPONSIBILITIES WITHIN A TRUSTWORTHINESS PROGRAMME FOR NUCLEAR SECURITY

3.1. Roles within a trustworthiness programme include those of the State, designated competent authorities, licensees, individuals and other stakeholders. Trustworthiness programmes can vary from one State to another. In some States, one competent authority might be responsible for conducting both trustworthiness assessments and determinations. In other States, the licensee might be responsible for both trustworthiness assessments and determinations. Still in other States, a combination of the two might exist. This section describes the case in which designated competent authorities conduct the assessments and determinations. However, the recommendations should be understood as applying to assessments and determinations conducted by either competent authorities, licensees or a combination of the two.

ROLE OF THE STATE

3.2. The State should establish a trustworthiness programme for nuclear security within the context of its legislative framework. A competent authority should be designated or established in State law to develop the regulatory requirements for the establishment and implementation of a trustworthiness programme for the State's nuclear security regime in order for licensees to grant access authorization to nuclear material and nuclear facilities, other radioactive material and associated facilities and activities, and/or sensitive information and sensitive information assets. The State should designate one or more competent authorities to be responsible for conducting or regulating and overseeing the trustworthiness programme. The State should provide to these competent authorities any necessary information on the individuals being assessed that is available only to the State.

3.3. The legal provisions of a trustworthiness programme for nuclear security should be consistent with national regulations pertaining to the collection and use of personal information to ensure an equitable and sufficiently robust system for making a trustworthiness determination. Depending on the State's legal framework, specific provisions might be established for the collection of personal information and the conduct of trustworthiness assessments. Additionally, consideration should be given to establishing a legal requirement for cooperation and coordination with other competent authorities to allow compliance with a request for the information needed to make a trustworthiness determination. Laws could also be established, unless conflicting with other legislation, to require that individuals disclose previously protected information. In such cases, legal precautions should be taken to ensure that such information is used only within the scope of the trustworthiness determination.

3.4. A State might already have an established trustworthiness programme that is not specific to its nuclear security regime. In this case, it might be appropriate to utilize previously established and authorized trustworthiness determinations made by other competent authorities (e.g. intelligence, law enforcement, first responders) for nuclear security purposes.

3.5.   A State might pass legislation or establish arrangements for the reciprocal recognition and communication of trustworthiness determinations between nuclear facilities and facilities using other radioactive material and relevant government agencies, including competent authorities.

3.6.   State provisions should exist to ensure the integrity and confidentiality of the assessment process and the outcome of trustworthiness determinations. Provisions for appropriate sanctions should be made, for example, to address falsification of, misrepresentation of or tampering with data or the process, either by the individual undergoing the trustworthiness assessment or by those responsible for the assessment or determination. Any falsified records, misrepresentation or tampering with data or the process should be rectified and recorded for future reference, possible root cause analysis and/or identification of trends.

Enforcement of the trustworthiness programme

3.7.   The State should establish sanctions to enforce the policies of the trustworthiness programme. Sanctions should be based on legal obligations that identify applicable offences and penalties related to the implementation of the trustworthiness programme. The State's legal framework can include sanctions for employees for mishandling, misuse or unauthorized actions in relation to sensitive information. Sanctions could also address unauthorized actions undertaken by organizations (e.g. at the level of the State, competent authority or licensee). Sanctions should be commensurate with the severity of the violation and could include suspension with or without pay, downgrade of the role of the employee, stripping of access to and authority over sensitive information, employee reassignment or termination of employment. States can also pursue civil or criminal sanctions.

ROLE OF THE COMPETENT AUTHORITY RESPONSIBLE FOR OVERSEEING A TRUSTWORTHINESS PROGRAMME AND MAKING TRUSTWORTHINESS DETERMINATIONS

3.8. The State should designate a competent authority to oversee the trustworthiness programme, supported by an appropriate legal framework.[1] The designated competent authority should develop regulations and guidelines for the trustworthiness programme, as well as oversight arrangements, data policies and procedures. The competent authority should ensure that the guidelines and procedures on trustworthiness are consistently applied by the licensee. This could be accomplished through an inspection programme (see paras 11.5–11.7).

3.9. Regulatory processes for trustworthiness assessments and determinations should document the following items, consistent with the State's legal framework:

(a) Description of the procedures for conducting pre-employment trustworthiness assessments and making trustworthiness determinations;

(b) Description of the procedures for performing periodic trustworthiness assessments and making new trustworthiness determinations during employment, including when changes are reported (e.g. those that are self-reported or reported by the licensee);

(c) Identification of types and sources of information that can be used to conduct trustworthiness assessments;

(d) Application of criteria and processes for validating or possibly mitigating derogatory information, circumstances or actions;

(e) Verification of the qualifications, experience and trustworthiness of those administering the processes;

(f) Confirmation of the period during which the determination will be considered valid;

(g) Arrangements for the management of personnel data using record keeping procedures to document completed trustworthiness assessments and determinations, including any documentation associated with appeals;

(h) Description of the procedures for protecting sensitive information gathered during the trustworthiness assessment;

(i) Application of provisions for the oversight of the trustworthiness programme;

(j) Description of the frequency of trustworthiness assessments and description of procedures for updating records.

---

[1] In some States, the licensee is responsible for trustworthiness determinations. In this case, the licensee should assume the roles and responsibilities related to trustworthiness determinations.

3.10. In some States, the competent authority responsible for oversight of the trustworthiness programme and the competent authority for the licensee are two different entities. However, in many States, the competent authority for the licensee also serves as the competent authority responsible for oversight of the trustworthiness programme and is responsible for making a trustworthiness determination. The designated competent authority should follow regulations and guidelines established for the trustworthiness programme and the established process for trustworthiness determinations. In addition, the designated competent authority should develop a notification process to communicate to the licensee in a timely manner any relevant information regarding an individual's trustworthiness determination, which could include information on issues that have arisen during the trustworthiness assessment. The licensee should notify the competent authority making the trustworthiness determination of relevant matters that come to its notice. In States where the licensee is responsible for either assessments, determinations or both, the competent authority responsible for the oversight of the trustworthiness programme should verify (e.g. through reviews) that the licensee is properly conducting assessments and/or making determinations.

Responsibilities of personnel making a trustworthiness determination

3.11. A trustworthiness determination should be made by suitably qualified and experienced personnel whose own trustworthiness has been verified, and it should be based on a balanced judgement of both the positive and negative aspects of the individual who is subject to the determination. The determination should be decided in a non-discriminatory manner, without personal bias, weighing the different factors and the security requirements of the job position in order to reach a conclusion on the individual's suitability to access nuclear material and nuclear facilities, other radioactive material and associated facilities and activities and/or sensitive information and sensitive information assets.

3.12. Trustworthiness determinations should be based on all of the information provided, and if that information is insufficient, further inquiries should be undertaken.

3.13. The personnel responsible for trustworthiness determinations should be aware of their State's legal framework concerning the protection of privacy and measures to ensure information security, including how information is processed within their own organization, how it is shared with external organizations and the penalties for the misuse of information or not properly protecting information.

3.14. The personnel responsible for trustworthiness determinations should not make a trustworthiness determination where there might be a conflict of interest in relation to the individual undergoing assessment; for example, if they know the individual in a personal capacity or if the person being assessed is responsible for their performance appraisal.

3.15. The personnel responsible for the trustworthiness determination are accountable for their decisions relating to an individual made at particular points in time, but the licensee should manage any subsequent issues that arise within the period of the trustworthiness determination.

ROLE OF THE COMPETENT AUTHORITY RESPONSIBLE FOR TRUSTWORTHINESS ASSESSMENTS

3.16. The State should designate a competent authority responsible for conducting trustworthiness assessments.[2] This competent authority should follow regulations and guidelines established for the trustworthiness programme and the established process for gathering information. The competent authority conducting the trustworthiness assessment may need to rely on other competent authorities within a State to supply personal information about the individual undergoing the trustworthiness assessment. This information could include, for example, criminal records, financial data, national residency data or historical records of professional conduct. Depending on the type of information needed for the trustworthiness assessment (see Section 4), more than one competent authority could be responsible for submitting information. Competent authorities should establish processes to preserve data privacy and information integrity and to transmit information in a secure and lawful manner.

Gathering information for trustworthiness assessments

3.17. The State should consider establishing regulations that specify other competent authorities (e.g. intelligence agencies, law enforcement agencies), third parties (e.g. lending institutions, previous employers, universities) and persons (e.g. relatives, friends, business associates) who can be approached to provide information on the individual undergoing assessment. The trustworthiness assessment process generally includes consent, through a physical or digital signature, from the individual under assessment.

---

[2] In some States, the licensee is responsible for trustworthiness assessments. In this case, the licensee should assume the roles and responsibilities related to trustworthiness assessments.

3.18. While the information provided by other competent authorities, third parties and individuals should generally be considered to be accurate, mistakes can nevertheless be made. Any discrepancy identified in the information provided should be verified before the trustworthiness determination is made. Such verifications should also be performed in the event of a negative trustworthiness determination, before the individual is notified.

3.19. The competent authority should request information to be submitted in a timely manner and should ensure that any sensitive personal information is received through a secure medium. Delays in receiving relevant information could delay employment decisions or necessitate expensive measures by the licensee (e.g. escorting) until such time as the trustworthiness determination is completed. Therefore, actions should be taken to avoid unnecessary delays in trustworthiness determinations.

ROLE OF LICENSEES

3.20. Licensees should be required to implement trustworthiness programmes as part of their licensing or other regulatory requirements. Licensees should maintain up-to-date records of employees, contractors and vendors who have undergone trustworthiness determinations, in accordance with national requirements.

3.21. The licensee should develop a list of the job positions subject to a trustworthiness determination using a graded approach and based on a job task analysis (see para. 4.7). This list should be submitted by the licensee to the competent authority for approval. Individuals seeking a position with the licensee should be informed of the trustworthiness assessment and determination process when they apply for employment.

3.22. Licensees should ensure that individuals who are subject to a trustworthiness determination are aware of their responsibilities within the trustworthiness programme, including reporting aberrant behaviour observed in others or changes in their own personal circumstances.

3.23. Licensees should have documented procedures for reporting trustworthiness programme data to the designated competent authority.

3.24. Licensees should follow approved procedures to collect information (e.g. identify documentation, work visa) during the pre-employment process prior to seeking a trustworthiness determination. The licensee should, where possible, maintain an effective working relationship with the competent authorities (e.g. the competent authorities responsible for trustworthiness programmes, intelligence agencies, law enforcement agencies) and with local law enforcement in accordance with State laws, regulations and policies.

3.25. While awaiting a trustworthiness determination for an individual, if a licensee identifies aberrant behaviour being exhibited by an individual, the behaviour should be reported to the competent authority making the trustworthiness determination. The licensee should nevertheless proceed with any disciplinary action and not wait for the determination from the competent authority. Trustworthiness determinations and performance management are separate processes, and thus licensees should not use trustworthiness determinations as a substitute for disciplinary action.

ROLE OF INDIVIDUALS SUBJECT TO TRUSTWORTHINESS DETERMINATIONS

3.26. Individuals subject to trustworthiness determinations should report the necessary personal information accurately, honestly and in a timely manner, during both the pre-employment and employment periods.

ROLE OF OTHER STAKEHOLDERS

3.27. Other organizations might employ individuals whose work duties fall within the framework of a State's nuclear security regime and therefore could also be subject to a trustworthiness determination by their respective competent authority. For example, law enforcement personnel, customs officials and border security personnel involved in systems and measures to detect and respond to material out of regulatory control can be subject to a trustworthiness programme by their respective competent authorities. In such cases, the relevant competent authorities should coordinate with each other to establish procedures to share information related to trustworthiness determinations.

3.28. Many States rely on personnel from organizations such as universities to provide scientific support to law enforcement agencies responding to incidents involving nuclear or other radioactive material out of regulatory control. In such cases, the State should consider having the personnel involved obtain a positive trustworthiness determination before being eligible to provide such support.

ACTIONS IN THE TRUSTWORTHINESS PROCESS

3.29. The trustworthiness assessment and determination process is initiated when an individual needs access to nuclear material and nuclear facilities, other radioactive material and associated facilities and activities, and/or sensitive information and sensitive information assets. In order for the licensee to grant access, the trustworthiness of the individual needs to be determined. Figure 1 illustrates the actions that the licensee and competent authorities take during the trustworthiness assessment and determination process in order to grant access to an individual.

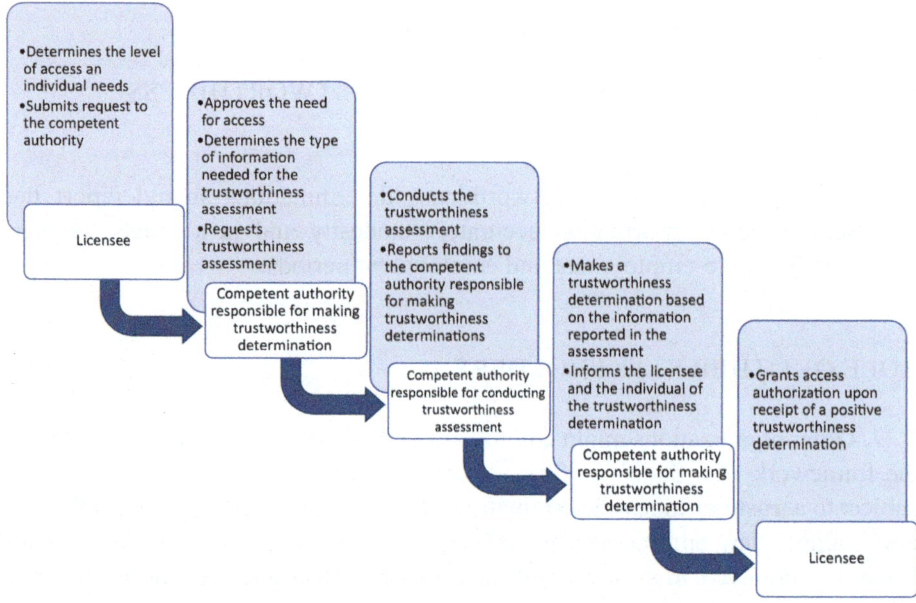

FIG. 1. *Actions of the licensee and competent authorities in the trustworthiness process. If the trustworthiness assessment or determination, or both, are the responsibility of the licensee, the actions described apply to the licensee.*

4. IMPLEMENTATION OF A TRUSTWORTHINESS PROGRAMME FOR NUCLEAR SECURITY USING A GRADED APPROACH

4.1. The designated competent authority should develop trustworthiness determination criteria to ensure a consistently applied and effective trustworthiness programme for nuclear security. These criteria should consider risk reduction, cost effectiveness and State laws for privacy as they relate to those individuals who are subject to a trustworthiness determination.

4.2. The State can establish effective and consistent requirements for trustworthiness determinations using the risk management objectives (e.g. potential consequences of unauthorized removal or sabotage) of the nuclear security programme [11].

4.3. Implementing all of the elements of a trustworthiness programme for all individuals involved in a nuclear programme could be challenging in terms of both cost and human resources. For this reason, a graded approach should be followed in the implementation of trustworthiness assessments.

4.4. As part of a graded approach, different types of information should be collected, in accordance with State laws, for the trustworthiness assessment on the basis of the attractiveness and category of the nuclear material, the attractiveness and category of other radioactive material, access to vital areas of the facility and/or the class of sensitive information, as appropriate. Examples of information that could be collected include the following:

(a) Identity verification (including dual nationality status);
(b) Previous employment verification, which could include reports or files related to personnel security;
(c) Education verification;
(d) Relevant personal history, which could include details concerning family members, co-residents, friends, colleagues and travel history;
(e) Financial evaluation;
(f) Criminal history review;
(g) Character evaluation;
(h) Social media review;
(i) Drug and alcohol testing;
(j) Medical information relevant to performing job functions;

(k)   Psychological assessment from a suitably qualified psychologist, for the most critical positions;

(l)   Polygraph testing, for the most critical positions.

4.5.   Table 1 presents an example of the information that could be used to assess the trustworthiness of an individual for access to each category of nuclear material and other radioactive material, to vital areas and to each class of sensitive information. The trustworthiness assessment criteria should be commensurate with the access authorization needed and the potential consequences of a criminal or intentional unauthorized act based on a risk informed approach [6]. The trustworthiness assessment criteria were developed using the categorization tables for nuclear material and other radioactive material provided in table 1 of Ref. [1] and table 7 of IAEA Nuclear Security Series No. 11-G (Rev. 1), Security of Radioactive Material in Use and Storage and of Associated Facilities [20], respectively, and the classification scheme for sensitive information described in Ref. [8].

4.6.   Trustworthiness assessments for individuals with access to the protected area, vital area, inner area or hardened room at a nuclear facility should be more frequent and comprehensive than trustworthiness assessments for individuals with access only to the limited access area. Individuals who participate in critical activities, such as the transport of nuclear material, should also undergo trustworthiness assessments. Figure 2 illustrates a typical layout at a nuclear facility. In such a facility, for example, a trustworthiness assessment for an individual who needs authorized access to the hardened room or enclosure is more comprehensive than a trustworthiness assessment for an individual who needs authorized access to the inner area or the vital area. Similarly, a trustworthiness assessment for an individual who needs authorized access to the inner area and the vital area is more comprehensive than a trustworthiness assessment for an individual who needs access to the protected area. All of the above areas involve more comprehensive assessments than an assessment undertaken for the limited access area.

4.7.   Once the attractiveness and category of the nuclear material, the attractiveness and category of other radioactive material, the vital areas of the facility and/or the class of sensitive information and any other assets that require access authorization at a facility have been identified, a job task analysis should be conducted. The job task analysis, which should consider access, authority and knowledge, is used to determine which job positions need direct access to secure areas and which can be limited to less secure areas. The facility management can then select a minimal number of individuals who will have authorized direct and unescorted or unsupervised access to nuclear material, other radioactive material, sensitive information and other equipment and systems. These are the individuals

TABLE 1. EXAMPLE OF A GRADED APPROACH TO TRUSTWORTHINESS ASSESSMENTS BASED ON THE CATEGORY OF NUCLEAR OR OTHER RADIOACTIVE MATERIAL, VITAL AREAS AND CLASS OF SENSITIVE INFORMATION TO WHICH AN INDIVIDUAL HAS ACCESS

Type of information collected[a]	Quantities below Category III nuclear material, natural uranium, depleted uranium and thorium, Category 4 and 5 radioactive sources	Category III nuclear material (limited access area), Category 1, 2 and 3 radioactive sources, restricted information[b]	Category II nuclear material (protected area), vital areas, confidential information[b]	Category I nuclear material (inner area), vital areas, secret information[b]
Identity verification	x	x	x	x
Employment verification	x	x	x	x
Education verification	x	x	x	x
Personal history		x	x	x
Financial evaluation		x	x	x
Criminal history		x	x	x
Character evaluation			x	x
Social media review	x	x	x	x
Drug and alcohol testing			x	x
Medical evaluation		x	x	x
Psychological assessment				x
Polygraph testing				x

[a] The type of information collected may differ from State to State.

[b] Classification of sensitive information depends on the national legal framework.

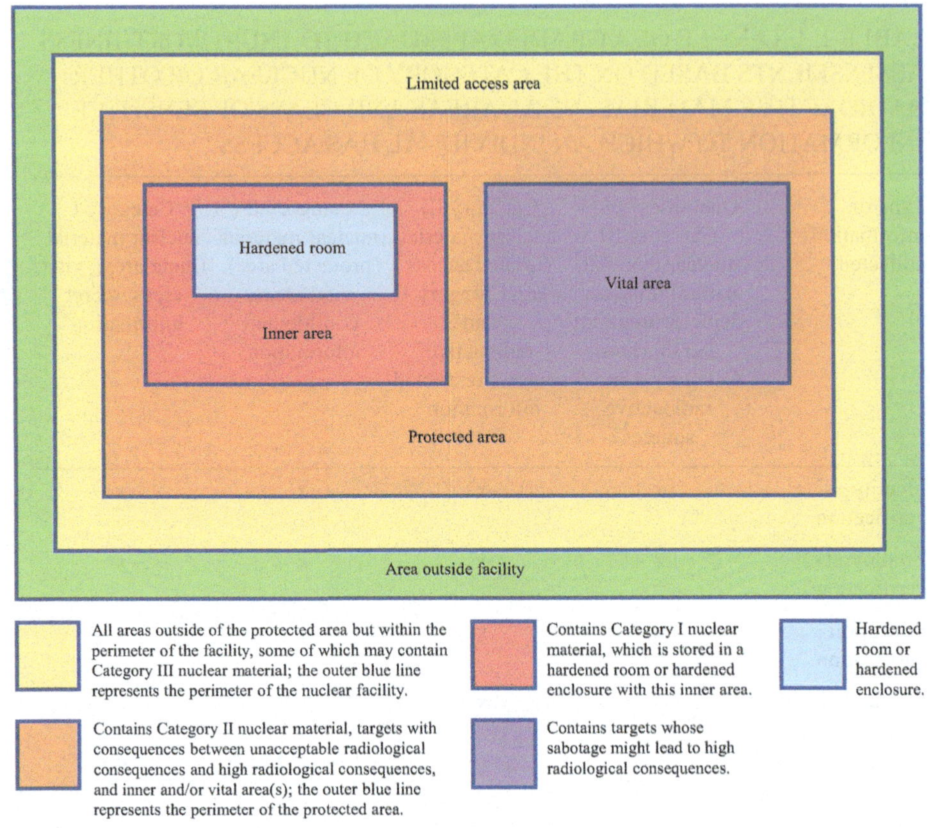

FIG. 2. Nuclear facility layout (adapted from fig. 7 of Ref. [11]).

for whom the most rigorous trustworthiness assessments should be conducted. The relevant job positions, as well as the individual's employment history and knowledge of the facility, should be taken into account when applying a graded approach. Examples of relevant job positions include nuclear material handlers, central alarm station operators, alarm technicians, reactor operators and engineers, maintenance engineers and technicians, personnel conducting vulnerability assessments, research scientists, information technology staff, other operations staff, and individuals involved in the transport of nuclear material. Depending on State laws, the individuals in these job positions are likely to be subject to ongoing trustworthiness assessments, drug and alcohol testing, and psychological personality profiling evaluations.

TRUSTWORTHINESS ASSESSMENTS OF OTHERS

4.8. Trustworthiness assessments should not be limited to employees of the licensee who are nationals of the country where the facility is located (either physically present at the facility or working remotely). Assessments should also be extended to foreign personnel, temporary personnel, students and interns, supply chain employees, contractors and visitors to the facility.

Foreign personnel

4.9. Foreign nationals support many nuclear activities, for example, training and education, maintenance during plant outages and decommissioning. Such foreign personnel may need access to nuclear material and nuclear facilities, other radioactive material and associated facilities and activities, and/or sensitive information and sensitive information assets, including those that can be accessed remotely, to conduct their assigned duties. Such arrangements should be evaluated closely before their finalization to ensure compliance with national laws and related organizational policies (e.g. with respect to export control and visa requirements). Once full compliance with national laws has been assured, host States could establish agreements (e.g. memoranda of understanding) with other States or international organizations to obtain information on the personal background and employment history of foreign nationals to assist in the trustworthiness determination process. Information provided by other States or international organizations should be verified to the extent possible. If personal information cannot be obtained from the State in question, access authorization or employment could be restricted or denied. If State laws permit, access could be granted through a special decision by the competent authority. In such cases, supplementary measures should be applied (e.g. escorting).

4.10. The legislation of a State determines whether arrangements allow for trustworthiness assessments of foreign nationals to be performed. Where they do, States should be aware that, if an individual has limited residency in the host State, there could be insufficient information available to support a comprehensive trustworthiness assessment. The provision of foreign police certificates could provide additional assurance of the trustworthiness of the individual, as could the verification of a trustworthiness determination in the individual's home State, if such verification arrangements exist. States should ensure that foreign personnel have the right to work in the host State and that trustworthiness determinations do not extend beyond the expiry of any work permits. Any behavioural indicators suggesting divided loyalty should lead to an additional trustworthiness assessment.

Temporary personnel

4.11. Paragraph 4.20 of Ref. [6] states:

"Persons whose trustworthiness has not been determined or whose duties do not require a trustworthiness assessment (e.g. temporary repair staff, administrative staff, maintenance staff, construction workers, visitors) should be escorted into vital areas or inner areas by persons who have authorized access and are not required to be themselves escorted."

Personnel without a trustworthiness determination should also be escorted into limited access areas, protected areas, and hardened rooms or enclosures.

4.12. A graded approach should be applied for trustworthiness assessments that might lead to escorted access of temporary workers. Depending on State laws and limitations (e.g. time, cost, human resources), trustworthiness assessments for temporary workers could be limited to verification of employment and identity, and criminal records.

4.13. Temporary workers who need access to nuclear material and nuclear facilities, other radioactive material and associated facilities and activities, and/or sensitive information and sensitive information assets should undergo a trustworthiness assessment based on a graded approach (see Table 1) to be granted an access authorization to perform their duties. These duties could include technical support activities, such as nuclear facility design, calibration and maintenance, operational and security functions, or computer security support functions. When the financial costs and time needed to make a trustworthiness determination associated with granting temporary access are high, escorted access should be considered.

4.14. When the temporary worker belongs to an external organization, appropriate arrangements should be made by the licensee to liaise with the external organization to ensure that an appropriate trustworthiness determination is transferred and that concerns are properly communicated to the licensee.

4.15. The licensee should ensure that temporary workers are treated in a manner equivalent to permanent personnel, although employment rights may differ. Given the transient nature of such temporary positions, arrangements to deny the access of temporary workers to the premises of the licensee or to sensitive information, including remote access, should be made in real time, either when the temporary worker leaves the licensee or when the contract ends.

4.16. The nuclear security culture should encourage personnel, contractors and temporary workers to feel equal responsibility for reporting concerns relating to the trustworthiness of individuals, despite the difference in employment status, even where it would necessitate reporting on other individuals. To maximize the likelihood of reporting, a reporting hotline should be made available to all personnel, whether temporary or permanent.

Students and interns

4.17. Depending on State laws and limitations (e.g. time, cost, human resources), trustworthiness assessments for students and interns could be limited to a verification of their identity, nationality and educational institution. Students and interns whose trustworthiness has not been determined should be escorted at all times when accessing an area for which a trustworthiness determination is necessary.

Supply chain employees

4.18. A large number of external organizations can be involved in the supply chain, either from other regions of the State or from other States. The globalization of the supply chain presents a number of challenges for competent authorities and licensees.

4.19. Personnel in the supply chain should feel equal responsibility for reporting aberrant behaviour or concerns relating to the trustworthiness of individuals, even where it would necessitate reporting on other personnel, contractors and temporary workers. To maximize the likelihood of reporting, any reporting hotline should also be made available to personnel in the supply chain.

4.20. Given the transient nature of supply chain employment at facilities, arrangements to deny access to the premises of the licensee or to sensitive information should be made in real time, either when the employee in the supply chain leaves the organization or when the facility's contract with the supply chain organization ends.

Information security

4.21. Individuals in the supply chain may have access to potentially sensitive information, including information on the design necessary for the production of specific items procured by the licensee.

4.22. Individuals within the supply chain who have direct or remote access to sensitive information should be subject to the same level of trustworthiness assessment as the staff of the licensee.

4.23. Contractual agreements can be established between States to extend the trustworthiness programme of the licensee to the personnel of a supply chain organization from another State. These agreements would enable the competent authority for the licensee to establish the requirements for the trustworthiness programme for organizations within the supply chain.

Counterfeit, fraudulent or suspect items

4.24. The increasing occurrence of counterfeit, fraudulent and suspect items arising within the global supply chain, including within the nuclear supply chain, is a cause for concern. These items often do not undergo the same rigorous quality control procedures as legitimate items and can deviate from prescribed specifications. Within the nuclear supply chain, counterfeit, fraudulent and/or suspect items can diminish the integrity of equipment, systems, structures, components or devices that contribute to nuclear safety and nuclear security.

4.25. The production and introduction of counterfeit and fraudulent items into the supply chain can be considered a criminal or intentional unauthorized act that can be perpetrated by entire companies or by individuals employed within the supply chain. The extension of a trustworthiness programme to organizations within the supply chain can thus assist such organizations in identifying individuals who are likely to participate in such acts.

4.26. Procedures should be implemented to determine the likelihood that a supply chain organization would willingly participate in the production and introduction of counterfeit, fraudulent and suspect items into the supply chain. Additional information on mitigating counterfeit, fraudulent and suspect items can be found in Refs [21, 22].

4.27. The actions taken by supply chain organizations to comply with the requirements of the trustworthiness programme for nuclear security contribute to enhancing the protection of sensitive information, mitigating counterfeit, fraudulent and suspect items, and increasing overall transparency and trust between the licensee and the entities within the supply chain.

Visitors

4.28. Depending on State laws and limitations (e.g. time, cost, human resources), trustworthiness assessments of visitors could be limited to verification of identity and nationality. Visitors whose trustworthiness has not been determined should be escorted at all times when accessing an area for which a trustworthiness determination is necessary.

4.29. States should consider the length of time needed to conduct a trustworthiness assessment for authorized access when they define visitors as either short term (i.e. when a trustworthiness assessment is unlikely to be needed) or long term (i.e. when a trustworthiness assessment should be completed based on a graded approach). The identity of all visitors should be verified regardless of the duration of the visit, as should the right of a foreign national to be present in the State.

4.30. If a visitor has not received a positive trustworthiness determination owing to either lack of information or a limited length of time for an assessment, the usual control measures, such as searching for recording devices or cameras, escorting and reporting of aberrant behaviour, should be implemented.

4.31. If a visitor is to be subject to a trustworthiness assessment, a statement that such an assessment will be performed should be included in the visitor request forms completed by the individual.

5. ESTABLISHING CRITERIA FOR CONDUCTING TRUSTWORTHINESS ASSESSMENTS AND MAKING TRUSTWORTHINESS DETERMINATIONS FOR NUCLEAR SECURITY

5.1. The frequency of trustworthiness assessments for nuclear security often differs from one State to another, since such assessments depend on national legislation; the regulatory approach used; the financial impact; the type of facility; the category of nuclear material; the category of other radioactive material and the associated facilities and activities; risk assessments; and the consequences of potential criminal or intentional unauthorized acts. The competent authority should define the minimum period between the trustworthiness determinations, while ensuring that trustworthiness assessments are supported by ongoing or continuous assessment on the basis of self-reporting or behaviour observation programmes. In States where the licensee is responsible for trustworthiness assessments and/or determinations, the minimum period between the trustworthiness determinations should be defined by the licensee, as appropriate, and subject to approval by the competent authority. The following have been identified as possible opportunities for the conduct of trustworthiness assessments:

(a)   During pre-employment periods;

(b)   Periodically during employment, including when there is a change in status or circumstances (e.g. new partner, new co-residents, conviction, debt, illness, trauma, extended absence, disciplinary proceedings, other significant life stressors);

(c)   After the occurrence of an incident that was defined on the basis of established criteria (e.g. inappropriate social media posting, suspicious approach from a third party, disciplinary action, positive test for alcohol or drugs);

(d)   Prior to assuming a new position that requires a more stringent trustworthiness assessment;

(e)   At the expiry date of a validity period for a trustworthiness determination.

DATA PRIVACY CONSIDERATIONS FOR TRUSTWORTHINESS ASSESSMENTS

5.2.   The collection and analysis of information in the context of trustworthiness assessments should be conducted in compliance with the State's legislation concerning the protection of privacy. Information from trustworthiness assessments should be considered sensitive, as this information could be used for blackmail or extortion. Most national regulations regarding privacy should therefore require the protection of this type of information [8]. Data should be protected in compliance with State laws. Legal, technical and administrative measures should be established and implemented to minimize intrusion into personal privacy and to maximize the protection of data, with the rules of the assessment process being made fully transparent and the individuals being informed of their rights regarding the protection of privacy and measures for information security. Such legal, technical and administrative measures should also address information sharing within the assessment process to facilitate decision making for trustworthiness determinations while also respecting individual privacy.

5.3.   Measures addressing information security should be implemented using a risk informed approach at all stages of the trustworthiness assessment process, for example during the collection, analysis, storage, retrieval, processing and destruction of information. Any access to documentation concerning trustworthiness assessments and determinations should be granted on a 'need to know' basis to authorized individuals to enable them to perform their assigned duties. States can assign classification levels to certain types of information in accordance with the guidance provided in Ref. [20].

BEHAVIOURAL INDICATORS OF INSIDER ADVERSARIES

5.4. Certain aberrant behaviours could indicate that an individual is predisposed to (or is vulnerable to being persuaded to) committing a criminal or intentional unauthorized act. While not all behaviour necessitates formal action, in combination with other factors, or when the consequences are significant and left unresolved, such behaviour could pose a risk to an individual's well-being, to the security arrangements of an organization or facility, or to the State's national security. For a strong nuclear security culture and to effectively reduce the risk of insider threats, both supervisors and co-workers have a responsibility to monitor and report any of these behavioural indicators observed in employees who are subject to a trustworthiness programme for nuclear security.

Aberrant behaviour

5.5. Certain aberrant behaviours could render an individual vulnerable to exploitation and could ultimately compromise an organization's security arrangements. The indicators listed below are more common than those associated with insider adversaries (see para. 5.6) but could indicate an increased potential for committing criminal or intentional unauthorized acts. Individuals experiencing stressful situations can sometimes exhibit aberrant behaviour that could result in a security concern. Where some doubt exists in relation to an individual's ongoing suitability to maintain a positive trustworthiness determination, early intervention is key to rapid and effective resolution. Examples of aberrant behaviour include the following:

(a) Use of illicit drugs and/or misuse of legal drugs or alcohol;
(b) Unwillingness to comply with rules, policies or procedures;
(c) Repeated irresponsibility in performing assigned duties;
(d) Disregard for authority or difficulty accepting feedback or criticism;
(e) Financial problems linked to gambling or financial irresponsibility;
(f) Indications of deceit, delinquent behaviour or lack of dependability;
(g) Symptoms of a mental or physical condition that impairs performance or adversely affects judgement;

(h)   Behaviour that results in criminal investigation, arrest or conviction, or legal infractions with a significant financial penalty;

(i)   Physical aggression or threat of destruction of property or life;

(j)   Difficulty dealing with stress or outwardly appearing to be under unusual stress;

(k)   Verbal hostility, aggression towards authority or co-workers, making malicious statements about co-workers and uncontrolled anger;

(l)   Significant changes in behaviour (e.g. moodiness, loss of inhibition, social disengagement);

(m)   Dissatisfaction with the employer or competent authority.

POTENTIAL INSIDER ADVERSARY INDICATORS

5.6.   If an individual exhibits one or more of the potential insider adversary indicators listed below, it could be cause for concern and merit further investigation. Insider adversary indicators could include the following:

(a)   Unexplained affluence;

(b)   Unreported conflicts of interest;

(c)   Failure to report foreign travel;

(d)   Unusual interest in information outside the current job scope;

(e)   Unusual work hours;

(f)   Unreported or concealed contacts with foreign nationals;

(g)   Unreported contacts with foreign governments or military or intelligence officials;

(h)   Attempts to gain access to sensitive information or sensitive information assets without a 'need to know' or authorization, including information on detection systems;

(i)   Unauthorized removal or sabotage of nuclear or other radioactive material, sensitive information or sensitive information assets, or sabotage of related facilities;

(j)   Attempts to gain access to areas for which access authorization has not been granted;

(k)   Attempts to provide access to persons for whom access authorization has not been granted;

(l)   Association or sympathy with criminal or terrorist individuals or groups;

(m)   Behaviour involving questionable judgement, integrity or organizational and national loyalties;

(n)   Uncharacteristic changes in appearance or behaviour;

(o)   Failure to disclose relevant medical conditions;

(p)   Unexplained absences.

ORGANIZATIONAL FACTORS ASSOCIATED WITH INSIDER ADVERSARIES

5.7. In some cases, organizational factors can develop over time and weaken the reliability and integrity of individuals. These organizational factors are a result of mismanagement rather than personal circumstances. Some examples of organizational factors that can contribute to a deterioration in the reliability and integrity of individuals over time include the following:

(a)   Poor communication by managers, leading to feelings of injustice;
(b)   Favouritism (e.g. rewarding or promoting less qualified individuals);
(c)   Failure by management to recognize exceptional efforts;
(d)   Poor treatment of the individual (e.g. harassment, lack of respect, hostile work environment);
(e)   Insistence on individuals working extra hours or in poor, stressful or hazardous work environments without compensation;
(f)   Dissatisfaction with salary, benefits or other forms of compensation;
(g)   Absence of, or lack of focus on sustaining, a nuclear security culture.

5.8. When making a trustworthiness determination, consideration should be given both to the number and combination of potential insider behavioural indicators displayed and to negative organizational factors that could suggest an increased potential to become an insider adversary.

5.9. Licensees should ensure that employment policies and contracts, including those in relation to the supply chain, support an environment that does not tolerate discrimination or unsatisfactory behaviour. There should be appropriate channels for raising concerns to reduce the possibility of individuals becoming disgruntled or disaffected as a result of unacceptable practices or of such practices not being properly addressed.

6. PROCESSES FOR THE IMPLEMENTATION OF A TRUSTWORTHINESS PROGRAMME FOR NUCLEAR SECURITY

6.1.   A trustworthiness programme for nuclear security should be implemented at various levels, including at the level of the State, competent authority and licensee, as well as at the level of other organizations supporting the functions of the licensee (e.g. contractors, vendors). The steps for implementing a trustworthiness programme are nevertheless the same for each level. Where the licensee is responsible for the trustworthiness assessment, the trustworthiness determination, or both, the described processes apply to the licensee.

6.2.   Trustworthiness programmes developed for other security areas of the State can be used in whole or in part for nuclear security. Alternatively, specific nuclear security trustworthiness programmes can be developed. Where a trustworthiness programme for nuclear security is built on existing State or industry programmes designed for other purposes, the competent authority should ensure that the objectives relating to nuclear security trustworthiness are consistent with and adequately incorporated into the objectives of the existing programmes.

6.3.   The following steps should be undertaken to establish a trustworthiness programme:

(a)   Designation of a competent authority;
(b)   Identification of stakeholders and a method of collaboration;
(c)   Development of governance and policy documents;
(d)   Implementation of training and awareness programmes;
(e)   Conduct of trustworthiness assessments;
(f)   Evaluation of the trustworthiness programme.

Designation of a competent authority

6.4.   The first step in implementing a trustworthiness programme is for the State to designate one or more competent authorities responsible for developing and implementing the programme. The designated competent authority should be familiar with relevant legislative and regulatory provisions and the role of a trustworthiness programme for nuclear security.

Identification of stakeholders and a method of collaboration

6.5. The implementation of an effective trustworthiness programme could involve collaboration by the competent authority with multiple stakeholders, including those who develop, manage, implement and are subject to the programme. Stakeholders can also include organizations with legal, medical, psychological and regulatory expertise, which can provide advice to the competent authority on ways to ensure that the programme meets the relevant requirements. Collaborative development of the programme with input from all these stakeholders provides mechanisms to identify issues or concerns early in the development process and supports the development of an effective and sustainable programme that meets the needs of all stakeholders.

Development of governance and policy documents

6.6. The State should provide guidance to the competent authority on the trustworthiness programme, including on privacy, and on the appropriate levels of transparency and effectiveness of the programme. This guidance should be established on the basis of legislative provisions and regulations. It should outline clear lines of authority and responsibility and describe the programme at a sufficient level of detail to ensure consistent implementation. The guidance should also identify oversight organizations; programme goals; organizations, activities and job positions subject to the trustworthiness programme; mechanisms by which individuals are assessed; and the process to appeal against a negative trustworthiness determination, if applicable. In accordance with State laws, guidance should also clearly communicate to those individuals subject to a trustworthiness determination: the requirements for their position; the information that will be gathered about individuals; the ways in which the information will be used, shared and stored; the obligations and rights of individuals with respect to the assessment process; and any arrangements to appeal against a negative trustworthiness determination, if applicable.

Implementation of training and awareness programmes

6.7. The competent authority responsible for conducting the trustworthiness assessment and the competent authority responsible for making trustworthiness determinations should receive training on how to conduct their tasks in accordance with the policies and procedures of the trustworthiness programme. The overall objective of the training programme is to ensure that individuals responsible for trustworthiness assessments and determinations are suitably qualified.

6.8.   Competent authorities should require licensees to raise awareness of the role that the trustworthiness programme has in the overall nuclear security culture, for example through a nuclear security awareness programme [6]. Individuals who are subject to trustworthiness assessments should be aware of their roles (e.g. reporting, following rules), the ways in which they can contribute to the process and the reasons why the programme is important for nuclear security (see paras 4.32–4.33 of Ref. [6]). Training and awareness programmes should be conducted on a regular basis to enhance commitment to the nuclear security culture.

Conduct of trustworthiness assessments

6.9.   Once guidance for the trustworthiness programme is established, and training and awareness activities are completed, the competent authority can initiate trustworthiness assessments.

Evaluation of the trustworthiness programme

6.10. Once a trustworthiness programme is implemented, competent authorities should periodically evaluate the programme. The evaluation should include audits and inspections of the trustworthiness programme and, where applicable, feedback from the licensee regarding the effectiveness of the programme. The results of the evaluation should be used to identify trends regarding trustworthiness related issues and to improve the programme. Guidance on audits and inspections of the trustworthiness programme is given in Section 11.

PROCESS FOR CONDUCTING TRUSTWORTHINESS ASSESSMENTS

6.11. Trustworthiness assessments can vary, but generally the following steps can be identified in the process for conducting trustworthiness assessments for nuclear security:

(1)   Determination of the access needed for the individual;
(2)   Submission of a request for a trustworthiness assessment;
(3)   Collection of information;
(4)   Validation of information and collection of additional information;
(5)   Determination of trustworthiness;
(6)   Documentation and communication of the trustworthiness determination;
(7)   Management of appeals.

Determination of the access needed for the individual

6.12. The licensee determines the level of access needed (e.g. limited access area, protected area, inner area, vital area; nuclear material or other radioactive material; type and class of sensitive information and sensitive information assets) for the individual on the basis of the individual's job functions and whether the position entails the individual working alone. Access levels can vary depending on the type of facility. The need for access should be confirmed by the manager or supervisor.

Submission of a request for a trustworthiness assessment

6.13. Once the access needed has been determined, the licensee submits a formal request to the competent authority to initiate the trustworthiness assessment. This request should be supported by an appropriate justification.

Collection of information

6.14. The competent authority responsible for trustworthiness determinations approves the assessment request and identifies the type of information needed to make a trustworthiness determination (see paras 6.19 and 6.20). The competent authority then requests the competent authority responsible for trustworthiness assessments to initiate the process of collecting and validating the information, ensuring that the information requests are in accordance with State laws, regulations and policies. In some cases, individuals might be requested to waive specific privacy rights so as to allow access to information that is necessary to conduct the assessment. These requests should be clearly communicated to the individual and should be considered only when necessary.

Validation of information and collection of additional information

6.15. After the information has been collected, the competent authority responsible for the trustworthiness assessment should validate the information provided. Criteria for what constitutes sufficient validation, and the methods used to validate information, should be documented in the procedures for trustworthiness assessments. For more comprehensive trustworthiness assessments, independent forms of validation from different organizations (e.g. government agencies, universities, financial institutions) could be needed for key pieces of information. During this comprehensive validation, the competent authority conducting the assessment should also gather any additional information needed to assess the individual against established criteria. Procedures should be defined to establish methods for collection of information, ensuring that such methods adhere to State laws, regulations and policies. Personnel responsible for gathering information should be suitably qualified and experienced. Records of all information gathering activities should be retained for audit purposes (see Section 11). A reporting process should also be established to allow individuals the opportunity to report inappropriate conduct by personnel gathering or validating information during the information gathering process.

Determination of trustworthiness

6.16. After the information has been gathered, the competent authority responsible for the trustworthiness determination assesses the information using an established methodology to determine whether the individual is suitable for a positive trustworthiness determination. Personnel responsible for making trustworthiness determinations should be suitably qualified and experienced in applying the methodology and in evaluating both aggravating and mitigating factors in a balanced manner to ensure that determinations are made consistently. Oversight of the personnel designated to make trustworthiness determinations should be established to ensure that one person is not solely responsible for trustworthiness determinations.

Documentation and communication of the trustworthiness determination

6.17. Trustworthiness determinations should be documented and communicated to the licensee that requested the determination. The trustworthiness determination should be communicated through secure and verifiable communication methods. Records of the trustworthiness determination, and justification for the corresponding access authorization, should be retained for potential audits of the evaluation process and for oversight purposes.

Management of appeals

6.18. Trustworthiness assessment processes could also include an appeal process, if permitted under State legislation, allowing individuals to appeal against a negative trustworthiness determination. Documentation should be made available on the appeal process and should include information on: what is considered acceptable grounds for appeal; the basis on which a trustworthiness determination can be reversed; designation of the authority granting or denying the appeal; the period within which a request to hear an appeal should be made; how the final determination is recorded and communicated; and whether any additional avenue of appeal exists (see paras 9.2–9.4 for more information on the appeal process).

SOURCES OF INFORMATION FOR A TRUSTWORTHINESS ASSESSMENT

6.19. Trustworthiness assessments involve the collection and review of personal information to justify the granting of a requested access authorization. The type of information used in the assessment should be determined by the competent authority and should be limited to the information necessary to evaluate an individual's trustworthiness using an established methodology. This information should be gathered in accordance with State laws and the protection of privacy.

6.20. Sources of information used in a trustworthiness assessment may include the following:

(a) Self-reported information and information reported by others. This can be obtained in written form (e.g. answers to a questionnaire). Interviews can also be used to resolve information gaps. Information gathered through questionnaires or interviews should be verified by independent sources, where possible. In addition, the identity of any person who provides information concerning an individual undergoing a trustworthiness assessment should be verified independently.

(b) Identity documents. Acceptable documents for the verification of an individual's identity include identification documents issued by the government or by other trusted organizations. Information that positively verifies the identity of the individual being assessed is an essential element of the trustworthiness assessment process.

(c) Personal record checks. These could include the verification of employment records, educational records, criminal history records or financial records. They may also include verification of other available information sources that relate to the established trustworthiness criteria.

(d) Physical examination records. Physical examinations could include drug and alcohol testing, medical evaluations, psychological assessments or polygraph testing.

(e) Publicly available information. This type of information, for example news articles, technical or scientific publications or presentations, self-published material, images, recordings, social media posts and information from artificial intelligence platforms, could also be consulted. When using publicly available information, special care should be taken to ensure data integrity and understand any bias that might be present.

(f) Information regarding abnormal activity in facility computer systems (e.g. physical protection systems, closed circuit television, nuclear material accounting and control systems).

FREQUENCY OF TRUSTWORTHINESS ASSESSMENTS

6.21. Trustworthiness assessments can be performed once, or they can be periodic, ongoing or incident based. The frequency of trustworthiness assessments should be determined by the competent authority using a graded approach on the basis of the individual's role and previous trustworthiness determinations.

6.22. Trustworthiness assessments are normally conducted once for individuals with temporary access or with limited access to nuclear material and nuclear facilities, other radioactive material and associated facilities and activities, and/or sensitive information and sensitive information assets.

6.23. Periodic trustworthiness assessments can be conducted at set intervals (e.g. every five years). The frequency of periodic assessments should be decided by the State or the competent authority. Periodic trustworthiness assessments could, for example, reveal relationships that are susceptible to collusion (e.g. marriage between individuals) or identify emerging factors (e.g. financial difficulties) that could serve as a source of coercion. For individuals with long term access to nuclear material and nuclear facilities, other radioactive material and associated facilities and activities, and/or sensitive information and sensitive information assets, periodic assessments are considered appropriate to assess whether the personal circumstances of an individual have changed and could potentially alter the initial trustworthiness determination. Additionally, the competent authority may implement continuous monitoring and/or flag changes to individuals' circumstances in the information or records systems.

6.24. Incident based trustworthiness assessments can be initiated on the basis of the results of reporting programmes or of relevant incidents. Such incidents could include self-reported status changes (e.g. criminal offences, serious financial issues, disciplinary actions, relevant psychological or other illnesses), security incidents or other criteria. The criteria to initiate a reassessment as a result of an incident should be documented by the competent authority as part of the procedures for the trustworthiness programme to ensure consistency and transparency when implementing the reassessment.

7. MAINTAINING RECORDS FOR A TRUSTWORTHINESS PROGRAMME FOR NUCLEAR SECURITY

TYPE OF DOCUMENTATION FOR A TRUSTWORTHINESS PROGRAMME

7.1. A trustworthiness assessment for nuclear security involves multiple types of documentation in paper or electronic format, such as the following:

(a) Requests, including justifications, by the licensee for trustworthiness assessments;
(b) Information provided by the individual undergoing the assessment in the form of answers to questions and verification documents;
(c) Information, verification documents and database queries or reports provided by third parties;
(d) The trustworthiness determination, including information concerning the justification for the trustworthiness determination by the competent authority;
(e) The basis for, and results of, any subsequent administrative actions (see para. 7.11).

7.2. The documentation for the trustworthiness assessment contains personal information, which should be subject to information security measures and made available on a strict 'need to know' basis. Information security measures protect the privacy of the individual and ensure that the individual is not exposed to attempts of coercion. Information should be shared only on a 'need to know' basis for each type of documentation in an effort to ensure that those initiating and conducting the trustworthiness assessments have access to only the information necessary to perform their assessment functions. Individuals who have access to view, create or modify documentation concerning trustworthiness assessments should themselves have previously undergone a trustworthiness assessment at the appropriate level, as determined by the State. When documentation concerning trustworthiness assessments is created or modified, it should be verified by a second person who is suitably qualified and has undergone a trustworthiness assessment at the appropriate level.

7.3. States should ensure that the processes associated with trustworthiness assessments are accompanied by a privacy statement indicating how personal information will be processed. This privacy statement should include the procedure to be followed in the case of a breach of these arrangements. Some States also have an official body to which breaches can be reported.

7.4. Personnel who manage or handle personal information should be appropriately trained and aware of the consequences of breaching the privacy statement, since such breaches could potentially undermine the willingness of individuals to disclose information relevant to the trustworthiness assessment and have reputational implications for the overall process and the licensee.

RECORD KEEPING FOR A TRUSTWORTHINESS PROGRAMME

7.5. Policies for record keeping should be established to ensure that records are retained for the competent authority to undertake audits of trustworthiness determinations (see Section 11). The need to preserve records for programme and evaluation purposes should be balanced with concerns associated with retaining large amounts of sensitive personal information. Personal information should be maintained for the period approved by the competent authority. The information used to make trustworthiness determinations and justifications for granting or revoking access authorization should be retained for the duration of the trustworthiness determination. Such information can also be maintained for a set period after the anticipated end of employment, or for longer periods where national security concerns have been identified. Evaluations should verify that policies and procedures are correctly followed in terms of record keeping, storage and disposal.

7.6. Policies for the disposal of records should be established to ensure that any documentation that is no longer needed is disposed of in a timely and secure manner to protect the privacy of individuals. Documenting the destruction of records can support the evaluation of record keeping arrangements.

7.7. Access to records associated with trustworthiness should be logged. Penalties and sanctions should be established and enforced for unauthorized access to, or use of, records associated with trustworthiness assessments to ensure that the information is used only for its intended purpose. Penalties and sanctions should be proportional to the potential consequences of the unauthorized access to, or use of, the records. Consideration should also be given to certain types of record (e.g. medical or psychological), for which the State might already have specific regulations in relation to record keeping.

7.8. A competent authority or licensee can be designated to oversee record keeping and to implement the temporary maintenance and subsequent destruction of trustworthiness documentation in accordance with State laws and regulations.

RECORDS OF ACCESS AUTHORIZATION

7.9. The licensee is responsible for maintaining a current list or database of approved access authorizations. This list should be subject to the appropriate information security measures to ensure the confidentiality, integrity and availability of information. The licensee should ensure that the list or database can be viewed only by individuals with a 'need to know' and that access to the list or database is logged. Only authorized individuals should have the ability to modify data. Measures to prevent the falsification of data by those who can view the list or database should also be implemented.

7.10. Licensees should grant access authorization for a designated period. The list or database of access authorizations should be continually reviewed and updated to reflect expiry dates of access authorizations, as well as to remove individuals who no longer have access authorization owing to changes in job functions, termination of employment, resignation or retirement, or changes in trustworthiness determinations. Individuals should be notified well in advance of the expiry of their access authorization to allow sufficient time for the submission of a renewal request and the supporting information, where necessary. Organizational changes, position changes and separation processes should be accompanied by notifications to the designated competent authority for the trustworthiness programme so that access authorization data can be updated accordingly and in a timely manner. Oversight evaluations should verify that access authorization data are recorded and are up to date.

RECORDS OF ACCESS SUSPENSIONS, DENIALS, REVOCATIONS OR RESTRICTIONS

7.11. The licensee is responsible for keeping records of suspensions, denials, revocations and restrictions of access authorizations for the period designated by the competent authority or State law. Such administrative actions can have implications for future applications for access authorization, and therefore these records should be maintained for future assessments. While managers should be informed of such administrative actions in order to implement the appropriate restrictions, the justifications for these actions (i.e. suspensions, denials, revocations or restrictions of access authorization) might include sensitive personal information. The continued protection of this information should remain of utmost importance, with the information shared only on a 'need to know' basis. Evaluations of the list or database of access authorizations should determine whether the administrative actions are implemented in a timely manner so that only those individuals appropriately authorized are able to maintain their access to nuclear material and nuclear facilities, other radioactive material and associated facilities and activities, and/or sensitive information and sensitive information assets.

8. GRANTING, SUSPENDING, REVOKING AND REINSTATING ACCESS AUTHORIZATION ON THE BASIS OF THE TRUSTWORTHINESS PROGRAMME FOR NUCLEAR SECURITY

GRANTING OF ACCESS AUTHORIZATION

8.1. Once the responsible entity within the licensee organization is notified that an individual has received a positive trustworthiness determination, the licensee can grant access authorization to the individual. Individuals who are granted positive trustworthiness determinations and receive access authorization are notified of their responsibilities and obligations under the trustworthiness programme for nuclear security. This notification should be given in a timely manner and should be documented in accordance with established procedures. The notification should provide the following information:

(a) The individual to whom the access authorization applies;
(b) The competent authority (or licensee in some cases) that has made the positive trustworthiness determination that allows the licensee to grant access authorization;
(c) The areas and information to which the individual has authorized access;
(d) The period for which the access authorization is valid;
(e) Any mandated reporting required by the State (e.g. change of partner or co-residents, financial issues, change of nationality, any new item on a criminal record) for the duration of the access authorization;
(f) Restrictions to access that may apply (e.g. medical conditions that restrict performing specific duties alone);

(g) A policy to comply with the access authorization conditions set by the licensee and, where relevant, the competent authority;

(h) Conditions under which the access authorization can be revoked.

8.2. The responsible entity within the licensee organization should be notified of any changes to an individual's trustworthiness determination in order to update the individual's access authorization.

Transfers of access authorization

8.3. On the basis of established procedures and agreements, the State and competent authority could allow the transfer of a positive trustworthiness determination between licensees in order to grant access to nuclear material and nuclear facilities, other radioactive material and associated facilities and activities, and/or sensitive information and sensitive information assets to temporary or permanent employees.

SUSPENSION OF ACCESS AUTHORIZATION

8.4. States should have a documented procedure to suspend a positive trustworthiness determination when needed. The suspension of a positive trustworthiness determination should result in the suspension of access authorization. If a positive trustworthiness determination is suspended, the individual concerned should be notified and informed in writing whether an appeal process is available (see paras 9.2–9.4).

8.5. Access control systems should be updated to reflect the revised status of the access authorization. The individual's supervisor or manager should also be informed, along with the human resources and medical departments (where appropriate), which can provide appropriate support until all avenues of appeal are exhausted, if applicable.

REVOCATION OF ACCESS AUTHORIZATION

8.6. States should have a documented procedure to revoke a positive trustworthiness determination where sufficient doubt exists in terms of the ongoing validity of the initial positive trustworthiness determination. The revocation of a positive trustworthiness determination should result in the revocation of access authorization. If a positive trustworthiness determination is revoked, the individual concerned should be notified and informed in writing whether an appeal process is available (see paras 9.2–9.4).

8.7. Access authorization is no longer needed when an individual leaves their employment or moves to a different position that does not need the same access authorization.

8.8. Access control systems should be updated to reflect the revocation of access authorization. If an individual's access was revoked as a result of a positive trustworthiness determination being revoked, the individual's supervisor or manager should be informed, along with the human resources and medical departments (where appropriate), which can provide appropriate support until all avenues of appeal are exhausted, if applicable.

REINSTATEMENT OF ACCESS AUTHORIZATION

8.9. The reinstatement of access authorization is possible for individuals who have been reassessed and received a positive trustworthiness determination (e.g. an employee who had left employment but has returned). The reinstatement of access authorization is also possible in the case that an individual has been successful in appealing against a decision to revoke a positive trustworthiness determination. Access control systems should be updated to reflect the reinstatement of access authorization.

9. PROTECTION OF INDIVIDUALS SUBJECT TO TRUSTWORTHINESS PROGRAMMES FOR NUCLEAR SECURITY

INFORMED CONSENT

9.1. A trustworthiness assessment programme for nuclear security should include an overarching statement outlining the purpose of the assessment and the means by which the assessment will be undertaken. This statement should make clear that a trustworthiness assessment is voluntary, but that failure by the individual to provide consent to undergo an assessment is likely to result in the withdrawal of an offer of employment. If the individual is already employed, refusal of consent to undergo an assessment could result in the termination of employment if access authorization is needed for the individual's job function. Where the individual provides consent to undergo a trustworthiness assessment, the individual should be made aware that they can withdraw consent at any time, although this might result in the termination of employment. Verbal consent should not be accepted; consent should be provided through a physical or digital signature.

APPEAL PROCESS

9.2. If State laws permit, States might wish to establish an appeal process to allow individuals the possibility of appealing against a negative trustworthiness determination or the revocation or suspension of a positive trustworthiness determination. The competent authority (or licensee in some cases) responsible for trustworthiness determinations should establish an appeal panel to schedule hearings, take records and report on the outcomes of hearings. The persons overseeing the appeal process should be independent of those who reached the negative trustworthiness determination.

9.3. The appeal process should afford the opportunity for the individual to attend a hearing in person, if they wish to do so. The appeal process should indicate whether individuals are permitted to be accompanied, for example by legal counsel, a union representative or management. At the conclusion of the hearing, the individual should receive a written decision outlining the grounds on which the decision was reached unless national security restrictions prevent this.

9.4. Establishing a transparent and independent appeal process helps to build confidence in the process and minimize the potential for legal challenges. Moreover, an appeal process enhances established standards for trustworthiness determinations by ensuring accountability.

10. CRITERIA FOR REPORTING WITHIN A TRUSTWORTHINESS PROGRAMME FOR NUCLEAR SECURITY

10.1. A key component to ensure the continuing trustworthiness of individuals is to foster a nuclear security culture that supports the monitoring and reporting of individuals' behaviour and actions and encourages self-reporting. Management in particular should be encouraged to observe the behaviour of personnel and report any behaviour of concern or non-compliance, with mandatory reporting to the independent department addressing trustworthiness concerns. The licensee should treat reporting as a legitimate and important function of the trustworthiness programme.

10.2. An effective trustworthiness programme for nuclear security should establish reporting procedures for individuals and relevant departments (e.g. human resources, medical, nuclear security). These procedures should include the following chain of command for reporting:

(1) Individuals report to management, to the human resources department or to the security department directly or anonymously (e.g. through a reporting hotline).
(2) The management, human resources or security department or hotline, as appropriate, reports to an independent department or interdepartmental multidisciplinary team that manages trustworthiness concerns.
(3) The independent department or interdepartmental multidisciplinary team that manages trustworthiness concerns conducts an inquiry.
(4) The results of the inquiry are reported to the competent authority or the entity within the licensee organization responsible for trustworthiness determination.

10.3. The competent authority (or licensee) determines whether a reassessment of the trustworthiness determination is necessary. The reporting procedures should provide clear criteria to help individuals exercise sound judgement on what and when to report. Reporting criteria can include indicators of aberrant behaviour that could potentially undermine nuclear safety and security.

10.4. The licensee should develop a training programme to effectively educate all individuals on their responsibilities in the context of the trustworthiness programme and on the overall criteria of the programme. Training should consider cultural barriers to reporting, with such barriers being addressed directly and attempts made to diminish them in order to ensure active engagement in the trustworthiness programme. Elements of the training programme should be included in induction arrangements (e.g. onboarding) and individuals should be subject to refresher training after a given period.

DIRECT OBSERVATION

10.5. An effective trustworthiness programme fosters a nuclear security culture that encourages individuals to be observant of behaviour or activities that could be an indication of insider adversary actions. Such actions could include the unauthorized removal or movement of nuclear material, tampering with the containment of nuclear material or falsification of records. For direct observation to be effective, individuals should be capable of recognizing unauthorized activities and should be able to properly report the activities to the appropriate personnel using the designated chain of command. Once reported, confirmation should be provided to the individuals that their report has been received and is being investigated. The outcome of the inquiry should be confidential.

SELF-REPORTING

10.6. An effective trustworthiness programme includes policies and procedures for individuals to self-report on factors that could merit further evaluation, as defined in the trustworthiness programme on the basis of the State's laws and regulations. Self-reporting could include reporting information on arrests or criminal charges, convictions, changes in name or personal status (e.g. marriage, divorce, birth or adoption of a child), foreign travel, ongoing foreign contacts, financial issues, bankruptcy, change in citizenship and medical issues including treatment for mental illness, as well as drug abuse or alcohol abuse.

10.7. The State or competent authority should establish the time frame for reporting, as well as the consequences of not reporting. The time frame for reporting personal changes could vary, depending on the nature of the information to be reported. For example, criminal or legal incidents might need to be reported in a more timely manner (depending on the nature of the offence) than a change in the individual's marital status.

REPORTING BY THIRD PARTIES

10.8. Part of an effective nuclear security culture is encouraging third parties (e.g. contractors, vendors, visitors) to be observant of any unusual behaviour or activities. Since third parties have their own chains of command, licensees should make arrangements for the management of vendors or contractors to be able to freely report to the licensee.

CONFIDENTIALITY AND NON-RETALIATION POLICIES

10.9. Confidentiality and non-retaliation policies are important aspects of a trustworthiness programme. Reporting procedures should include confidentiality and non-retaliation policies to ensure that individuals are treated fairly and that any information reported will remain confidential. These policies should also ensure protection against retaliation in cases of self-reporting, reporting behaviour of concern in good faith or participating in an inquiry.

10.10. The policy on non-retaliation should also stipulate actions that should be taken if concerns reported in relation to an individual's behaviour are deliberately false. The intended function of the legitimate activities that comprise the trustworthiness programme can be subverted by unscrupulous individuals for improper reasons. Such individuals can falsely accuse personnel or hide the criminal or intentional unauthorized acts of other personnel. If not appropriately addressed, false reports or failures to report inappropriate behaviour, can have a negative impact on the overall nuclear security culture.

PROCESS FOR CONDUCTING INQUIRIES

10.11. An inquiry process is necessary to examine reported information and to determine the appropriate response. The objectives of an inquiry are to gather information and facts, provide an opportunity for the individual under investigation to give input, and identify motivational or behavioural indicators (e.g. stress, financial irregularities, conflict, other workplace difficulties) in order to decide the appropriate response (e.g. report to the competent authority for reassessment, take employment actions, terminate employment). Inquiries should be impartial, thorough and completed in a timely manner. If not conducted correctly or fairly, inquiries can negatively impact morale and trust among personnel, reducing the effectiveness and overall benefit of the trustworthiness programme for nuclear security. The inquiry process could include the following steps:

(1) Determining the need for an inquiry;
(2) Planning the inquiry;
(3) Conducting the inquiry;
(4) Reporting the inquiry results;
(5) Taking appropriate action.

Determining the need for an inquiry

10.12.   The first step in the inquiry process is to decide whether an inquiry is needed, which can be determined on the grounds of the information reported. The licensee determines the need for an inquiry on the basis of a number of considerations, such as whether the information reported meets the defined criteria of a behaviour of concern. When the licensee receives reported information, the competent authority (or entity responsible for trustworthiness determination within the licensee organization) should be notified within the established time frame for the type of information reported. In some cases (e.g. reports of unauthorized removal of nuclear material, sabotage, misuse of sensitive information) the competent authority may decide to take over the inquiry.

Planning the inquiry

10.13.   Effective planning ensures that the inquiry is properly conducted. The licensee should determine the objectives and scope of the inquiry, the individual who will conduct the inquiry, the inquiry process and the time frame. The objectives and scope determine the activities of the inquiry. If the inquiry reveals issues that are additional to those originally reported, the individual conducting the inquiry could decide to include these additional issues and expand the scope of the inquiry, to initiate a separate inquiry or to disregard these other issues and focus on the initial scope of the inquiry. Individuals appointed to conduct inquiries should be impartial and possess the necessary skills to conduct such inquiries. These skills should be judged on the basis of the individual's level of training, education and experience.

Conducting the inquiry

10.14.   The individual assigned to conduct the inquiry should gather the relevant evidence and consider the nature of the report or concerns, in compliance with State laws and policies regarding privacy and access to information. The process can include interviews and document reviews (e.g. personnel files, records, logs, digital information, emails, text messages, computer files). Interviews of individuals under investigation should include a discussion on cooperation, confidentiality procedures and the importance of providing truthful and honest information. Interviews of individuals in support of the investigation should include information on non-retaliation policies, and on whether the individual under investigation can access the details of the interview. Individuals under investigation should be provided with support through an assistance programme from the moment they are notified that they are subject to an inquiry.

Reporting the inquiry results

10.15. The report on the inquiry results should contain factual conclusions that directly address the objectives of the inquiry, supported by the information gathered. The report should document the process, the information collected, the final determination and the justification for the conclusions.

Taking appropriate action

10.16. Management should take appropriate action on the basis of the final determination of the inquiry. Such action could include the following:

(a) Allowing the individual to maintain their positive trustworthiness determination while management monitors the individual's behaviour and work performance;
(b) Revoking or reducing the individual's access authorization while management addresses the causes of the behavioural issues and monitors the individual's work performance;
(c) Revoking the individual's access authorization and making a request to revoke the individual's positive trustworthiness determination;
(d) Taking no action, if concerns are determined to be unsubstantiated, inaccurate or misunderstood, or if there is a valid explanation for the reported behavioural concerns.

11. AUDITS AND INSPECTIONS OF A TRUSTWORTHINESS PROGRAMME FOR NUCLEAR SECURITY

AUDITS OF TRUSTWORTHINESS ASSESSMENTS AND DETERMINATIONS

11.1. Trustworthiness assessments and trustworthiness determinations should be subject to audit to confirm that the appropriate assessment and determination processes were followed. Audits should be done on a predetermined sample size, with cases selected at random.

11.2. Audits should consider whether trustworthiness assessments and determinations were conducted by suitably qualified and experienced personnel. Personnel that conduct trustworthiness assessments and determinations could be subject to bribes, which could result in the individuals receiving false positive trustworthiness determinations. Access authorizations might therefore be granted in cases where they otherwise would not, increasing the potential for criminal or intentional unauthorized acts by insider adversaries. Such issues could be mitigated through periodic or random audits of trustworthiness assessments and determinations, as well as through periodic trustworthiness reassessments of the individuals conducting assessments and determinations.

11.3. Audits can take the form of reviews conducted by a separate internal audit section of the competent authority or reviews conducted by an independent or national audit authority, including organizations with regulatory oversight. If the review is performed by an internal audit section of the competent authority, the personnel involved should be independent from those who conducted the original trustworthiness assessment or made the trustworthiness determination. Audits should not be conducted by the supervisors of the individuals who conducted the original trustworthiness assessment or made the trustworthiness determination.

11.4. The licensee should be notified when a trustworthiness determination is under audit. If the trustworthiness determination is found to have been made on the basis of an inaccurate assessment, a new assessment should be conducted. If a positive trustworthiness determination is revoked upon reassessment, individuals should be informed in writing of the reasons for the decision (see paras 8.6–8.8).

REGULATORY INSPECTIONS OF A TRUSTWORTHINESS PROGRAMME

11.5. To assess the trustworthiness programme for nuclear security established by the licensee, the competent authority for the licensee should conduct inspections at the premises of the licensee to discuss procedures with the personnel responsible for implementing the trustworthiness programme and to receive feedback from individuals selected at random, in relation to the trustworthiness programme. The results of such inspections should confirm that the licensee is implementing the trustworthiness programme appropriately.

11.6. Some inspections can be conducted remotely. For example, reviews of licensee policies (e.g. for social media, drug and alcohol testing, harassment and bullying) may be conclusive enough to avoid the need for an on-site visit to the licensee.

11.7. All inspections should conclude with the production of a written report that not only provides an overall view of the trustworthiness programme but also indicates areas that need improvement, along with good practices.

REFERENCES

[1]   INTERNATIONAL ATOMIC ENERGY AGENCY, Nuclear Security Recommendations on Physical Protection of Nuclear Material and Nuclear Facilities (INFCIRC/225/Revision 5), IAEA Nuclear Security Series No. 13, IAEA, Vienna (2011),
https://doi.org/10.61092/iaea.ko2c-dc4q

[2]   INTERNATIONAL ATOMIC ENERGY AGENCY, Nuclear Security Recommendations on Radioactive Material and Associated Facilities, IAEA Nuclear Security Series No. 14, IAEA, Vienna (2011).

[3]   EUROPEAN POLICE OFFICE, INTERNATIONAL ATOMIC ENERGY AGENCY, INTERNATIONAL CIVIL AVIATION ORGANIZATION, INTERNATIONAL CRIMINAL POLICE ORGANIZATION–INTERPOL, UNITED NATIONS INTERREGIONAL CRIME AND JUSTICE RESEARCH INSTITUTE, UNITED NATIONS OFFICE ON DRUGS AND CRIME, WORLD CUSTOMS ORGANIZATION, Nuclear Security Recommendations on Nuclear and Other Radioactive Material out of Regulatory Control, IAEA Nuclear Security Series No. 15, IAEA, Vienna (2011).

[4]   Amendment to the Convention on the Physical Protection of Nuclear Material, INFCIRC/274/Rev.1/Mod. 1 (Corrected), IAEA, Vienna (2021).

[5]   INTERNATIONAL ATOMIC ENERGY AGENCY, Nuclear Security Culture, IAEA Nuclear Security Series No. 7, IAEA, Vienna (2008).

[6]   INTERNATIONAL ATOMIC ENERGY AGENCY, Preventive and Protective Measures Against Insider Threats, IAEA Nuclear Security Series No. 8-G (Rev. 1), IAEA, Vienna (2020).

[7]   INTERNATIONAL ATOMIC ENERGY AGENCY, Nuclear Security Systems and Measures for the Detection of Nuclear and Other Radioactive Material out of Regulatory Control, IAEA Nuclear Security Series No. 21, IAEA, Vienna (2013).

[8]   INTERNATIONAL ATOMIC ENERGY AGENCY, Security of Sensitive Information for Nuclear Security, IAEA Nuclear Security Series No. 23-G (Rev. 1), IAEA, Vienna (in preparation).

[9]   INTERNATIONAL ATOMIC ENERGY AGENCY, Use of Nuclear Material Accounting and Control for Nuclear Security Purposes at Facilities, IAEA Nuclear Security Series No. 25-G, IAEA, Vienna (2015).

[10]   INTERNATIONAL ATOMIC ENERGY AGENCY, Security of Nuclear Material in Transport, IAEA Nuclear Security Series No. 26-G, IAEA, Vienna (2015).

[11]   INTERNATIONAL ATOMIC ENERGY AGENCY, Physical Protection of Nuclear Material and Nuclear Facilities (Implementation of INFCIRC/225/Revision 5), IAEA Nuclear Security Series No. 27-G, IAEA, Vienna (2018).

[12]   INTERNATIONAL ATOMIC ENERGY AGENCY, Preventive Measures for Nuclear and Other Radioactive Material out of Regulatory Control, IAEA Nuclear Security Series No. 36-G, IAEA, Vienna (2019).

[13] INTERNATIONAL ATOMIC ENERGY AGENCY, Security Management of Radioactive Material in Use and Storage and of Associated Facilities, IAEA Nuclear Security Series No. 43-T, IAEA, Vienna (2022).

[14] INTERNATIONAL ATOMIC ENERGY AGENCY, Self-assessment of Nuclear Security Culture in Facilities and Activities, IAEA Nuclear Security Series No. 28-T, IAEA, Vienna (2017).

[15] INTERNATIONAL ATOMIC ENERGY AGENCY, Enhancing Nuclear Security Culture in Organizations Associated with Nuclear and Other Radioactive Material, IAEA Nuclear Security Series No. 38-T, IAEA, Vienna (2021).

[16] INTERNATIONAL ATOMIC ENERGY AGENCY, Safety of Nuclear Power Plants: Commissioning and Operation, IAEA Safety Standards Series No. SSR-2/2 (Rev. 1), IAEA, Vienna (2016).

[17] INTERNATIONAL ATOMIC ENERGY AGENCY, Recruitment, Qualification and Training of Personnel for Nuclear Power Plants, IAEA Safety Standards Series No. SSG-75, IAEA, Vienna (2022).

[18] INTERNATIONAL ATOMIC ENERGY AGENCY, The Operating Organization and the Recruitment, Training and Qualification of Personnel for Research Reactors, IAEA Safety Standards Series No. SSG-84, IAEA, Vienna (2023).

[19] INTERNATIONAL ATOMIC ENERGY AGENCY, Assessing Behavioural Competencies of Employees in Nuclear Facilities, IAEA-TECDOC-1917, IAEA, Vienna (2020).

[20] INTERNATIONAL ATOMIC ENERGY AGENCY, Security of Radioactive Material in Use and Storage and of Associated Facilities, IAEA Nuclear Security Series No. 11-G (Rev. 1), IAEA, Vienna (2019).

[21] INTERNATIONAL ATOMIC ENERGY AGENCY, Managing Counterfeit and Fraudulent Items in the Nuclear Industry, IAEA Nuclear Energy Series No. NP-T-3.26, IAEA, Vienna (2019).

[22] INTERNATIONAL ATOMIC ENERGY AGENCY, Procurement Engineering and Supply Chain Guidelines in Support of Operation and Maintenance of Nuclear Facilities, IAEA Nuclear Energy Series No. NP-T-3.21, IAEA, Vienna (2016).

CONTACT IAEA PUBLISHING

Feedback on IAEA publications may be given via the on-line form available at:
www.iaea.org/publications/feedback

This form may also be used to report safety issues or environmental queries concerning IAEA publications.

Alternatively, contact IAEA Publishing:

Publishing Section
International Atomic Energy Agency
Vienna International Centre, PO Box 100, 1400 Vienna, Austria
Telephone: +43 1 2600 22529 or 22530
Email: sales.publications@iaea.org
www.iaea.org/publications

Priced and unpriced IAEA publications may be ordered directly from the IAEA.

ORDERING LOCALLY

Priced IAEA publications may be purchased from regional distributors and from major local booksellers.

Printed and bound by CPI Group (UK) Ltd, Croydon, CR0 4YY

06/07/2026

02160596-0001